AF697607

DISCOVERING

# DISCOVERING CYCLES

BY GLENN O. BLOUGH

PICTURES BY JEANNE BENDICK

In our world many things happen again and again. Each day the sun rises and sets, each year brings summer and winter, each autumn birds fly south and in the spring north. All around us there are cycles taking place.

This book describes the most important ones in our life: the living cycles of plants and animals; the changing seasons; the coming of rainy days and sunny days; the transfer of oxygen and carbon dioxide in plants and animals, and other cycles as well.

There are a variety of projects—from watching birds fly away in the fall to keeping aquariums. How we get day and night and the changing of seasons are explained. Readers can plant flowers, collect their seeds and plant new flowers to discover the life cycle. There are ways of condensing water and seeing how it rains and also how water evaporates again.

*(continued on back flap)*

*(continued from front flap)*

The readers will discover simple but ısic science concepts. In addition, they ill learn how to observe nature around em, and learn to discover the marvelıs world of cycles.

The lively, enjoyable illustrations of anne Bendick accompany the text. hey help the reader visualize the ideas esented in the book.

Dr. Glenn O. Blough is well known roughout the country as a leading ıthor and lecturer in the field of sciıce education. Formerly a specialist Elementary Science in the U.S. Office Education, he is now Professor of ducation at the University of Maryland. he 1958 President of the National Sciıce Teachers Association, Dr. Blough ıs written many books for young ople.

Jeanne Bendick is something of a ıenomenon among authors and illusators. In addition to the books she ıs written for very young readers, she ıs also written HOW MUCH AND OW MANY, ELECTRONICS FOR OUNG PEOPLE, and collaborated th her husband, Robert Bendick, on ELEVISION WORKS LIKE THIS ıd FILMING WORKS LIKE THIS. ıey live in Rye, New York.

*To my friend and teacher,*
*Gerald S. Craig* G.O.B.

To Rachel
J.B.

Copyright © 1973 by Glenn O. Blough and Jeanne Bendick. All Rights Reserved. Printed in the United States of America. No part of this publication may be reproduced, stored in a retrieval system, or transmitted, in any form or by any means, electronic, mechanical, photocopying, recording, or otherwise, without the prior written permission of the publisher.

**Library of Congress Cataloging in Publication Data**

Blough, Glenn O.
Discovering cycles.

SUMMARY: Text and suggested projects explore some of the cycles found in nature, such as day and night, the seasons, the water cycle, the cycle of growth, and others.
1. Biology—Periodicity—Juvenile literature. [1. Periodicity. 2. Nature study] I. Bendick, Jeanne, illus. II. Title
QH527.B56 574.1 73-6622
ISBN 0-07-006165-3

McGraw-Hill Book Company

*New York St. Louis San Francisco Düsseldorf Johannesburg Kuala Lumpur London Mexico Montreal New Delhi Panama Rio de Janeiro Singapore Sydney Toronto*

# DISCOVERING CYCLES

BY GLENN O. BLOUGH

PICTURES BY JEANNE BENDICK

59923

J.B.U. Library
Siloam Springs, Ark.

# Cycles

Every clear morning if you look toward the East, you will see the sun come up. Day comes. In the evening if you look toward the West, you will see the sun go down. Night comes. The next morning the sun will come up again and in the evening it will go down. Another day and another night.

Over and over again.

In March spring comes. That makes many people happy. In June summer comes. That makes people warm. In September it's autumn. School starts. People gather harvests and get ready for winter. In December winter comes. That makes people cold. Every year this happens. Spring, summer, autumn, and winter.

Over and over again.

Some days it rains. Water falls out of the air. Then it stops raining. The sun shines.

Water goes back into the air. Water comes out of the air. Water goes back into the air.

Over and over again.

These are examples of cycles—a cycle is something that starts, goes on, ends, and starts all over again. Wherever you live there are cycles. This is a book about some of the cycles you can discover around you. Some are short. Some take a long time.

Find a watch that has a second hand. Keep your eye on a starting place. Watch the hand. In sixty seconds it will be back where it started. Some cycles like this take only sixty seconds, or one minute. Watch the minute hand. How long does it take to make a cycle? How about the hour hand?

When you are discovering cycles you will use some of the ways a scientist uses to make discoveries. How did you find out about the watch cycles? You looked carefully, you *observed*. You *thought* about what you saw.

A magnifying glass will come in handy to help you discover about some of the cycles in this book. You may *experiment* to find out about others. When you are discovering cycles you may make drawings or keep records of the things you find out. Scientists often do that.

You will be surprised at the cycles that are going on around you.

# The Cycle of Days and Nights

It's night. You lie in bed. You look around and you cannot see the ceiling or the walls of your bedroom. You look out your window. You cannot see the trees or your neighbor's house or the sky. You may see a distant light or two and some stars. But it's dark.

Wait. It is time for sleep.

When you wake you can begin to see the ceiling and the walls of your bedroom. Outdoors you can see the trees and the houses and the sky. It is light.

Wait. It is time for school.

A day passes. Look out your window. It begins to get dark again. It gets darker and darker. Soon you can see a light or two and a few stars. It's dark again. You have watched the day and night cycle.

Once no one knew what made this cycle of day and night. People thought up all kinds of ways to explain it. Some people thought that the sun was a god that got up in the morning and drove a bright chariot across the sky. Then at night he put the horses and chariot away and went to bed.

But people began to observe the earth and the sun and to think about them. Now we know what makes the cycle of day and night.

The earth you live on is round like a ball, but when you look at it you can't tell that it is round. The earth is so big you can't see its roundness. You can fly high in an airplane and see for miles and miles, but you still can't see the earth's roundness. If you could fly toward the moon with the astronauts and look back at the earth, you would be able to see that it is round like a ball.

The earth you live on spins like a top. You can't tell that it is spinning because you spin with it. There wouldn't be a day and night cycle if the earth didn't keep spinning.

The sun shines on the earth you live on. This makes it light. If the sun didn't shine on the earth, we would always have night and there wouldn't be a day and night cycle.

You can understand the cause of day and night if you use a globe to represent the earth and a flashlight to represent the sun.

Find the place where you live on the globe and mark it. Remember that when this place moves you move.

Make the room as dark as possible and shine the flashlight on the globe. Now turn the globe around, but not in the same direction that the hands of the clock turn. Turn it the other way. We call this direction *counterclockwise*. Keep turning the globe until the place where you live comes back to where it started. This didn't take long for you to do, but do you know how long it takes the earth you live on to make a cycle like this? It takes 24 hours. If you watch the hour hand of a clock as the earth makes a cycle, you will discover that the hour hand makes two cycles. Each cycle takes 12 hours.

Some of these 24 hours are daylight. Some are dark. Use the globe and the flashlight and you will see why.

Have someone hold the flashlight (the sun) while you turn the globe (earth). Keep the flashlight shining on the globe. You can see that only half of the earth is lighted at one time. That's because the earth is round like a ball.

When you go home from school, play outdoors, come indoors for supper, and finally go to bed, the earth has been turning and turning. Now if you look outdoors, it will be dark again. Night is here.

START WITH THE SPOT TURNED AWAY.
NOW IT IS NIGHT WHERE YOU LIVE.

WHEN THE SUN IS HIGHEST
IT IS ABOUT NOON WHERE YOU LIVE.

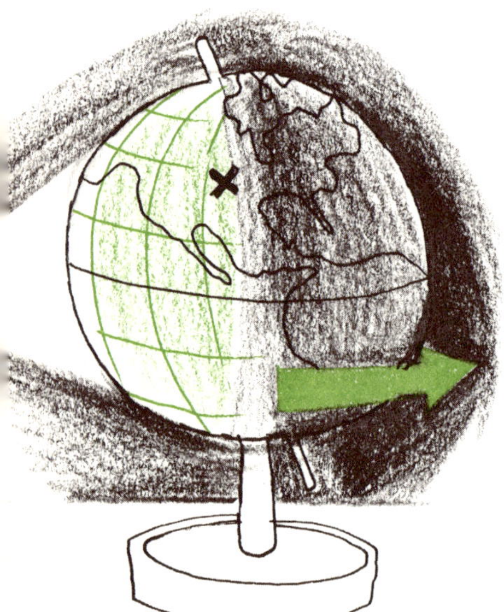

WHEN THE EARTH AND SUN ARE LIKE
THIS, IT IS ALMOST NIGHT AGAIN.

The earth keeps turning and the place where you live comes back to where it started. It has finished one cycle. We say the earth has *rotated* once. That means that it has spun around once. The cycle continues. Each time the earth rotates you are a day older. The earth keeps rotating and rotating—day and night, day and night, over and over again.

Remember that when the place where you live is turned toward the sun it is day. When it is turned away from the sun it is night. Remember that you did not move the flashlight (the sun). You moved the globe (the earth). The earth does the moving.

Does the sun really come up and go down? No, it only seems to do this. When the earth rotates that makes it look as if the sun were coming up and going down.

# The Cycle of the Seasons

You know that spring, summer, autumn, and winter are the seasons. They follow one after the other in a cycle. The four seasons make a year.

Summer days are warm. Winter days are cold. Spring and autumn days are in-between. They are not as warm or as cold as the other seasons.

No matter where you live in the United States there is a cycle of four seasons. They are not alike everywhere. In the northern parts of the United States the winters are very cold. In the southern parts they are warmer.

Once no one knew the cause of the cycle of the seasons. Now we know that there are two causes for the cycle. One is the way the earth moves, and the other is the way it is tipped.

You remember that the earth rotates every 24 hours. The earth moves in another way too. It moves around the sun. This is called *revolving*. The earth revolves around the sun once every year.

You are a year older each time the earth revolves. How many times has the earth revolved around the sun since you were born? If you are ten years old, it has revolved ten times.

The earth keeps rotating as it revolves. It rotates 365¼ times while it revolves once. That is the number of days in a year. Can you figure out how many times the earth has rotated since you were born?

The revolving of the earth is one of the very important reasons for the cycle of the seasons.

The way the earth is tipped is the other very important reason for the seasons.

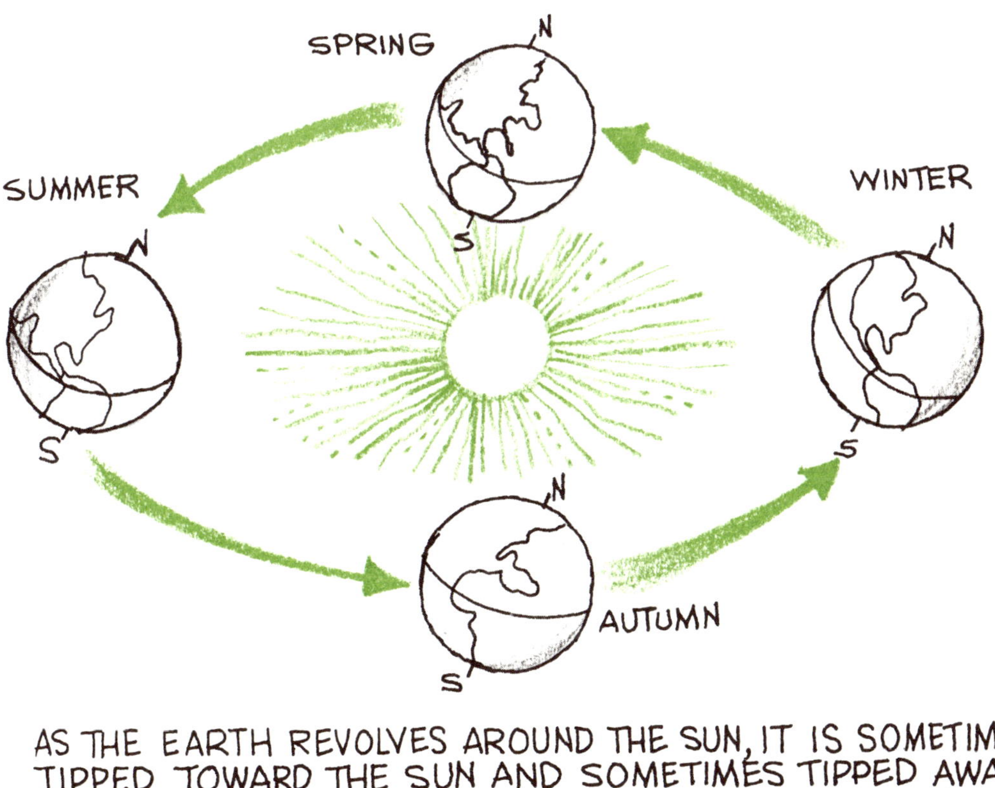


AS THE EARTH REVOLVES AROUND THE SUN, IT IS SOMETIMES TIPPED TOWARD THE SUN AND SOMETIMES TIPPED AWAY FROM IT. WHEN IT IS TIPPED TOWARD THE SUN, THE DAYS ARE WARMER. WHEN IT IS TIPPED AWAY, THE DAYS ARE COLDER.

Let's start our discovery of the cycle of the seasons by looking at summer days. You remember that the sun gives us light and heat. In summer it is probably daylight when you wake up in the morning, and it is probably light long after suppertime and you can play outdoors. When the sun shines on the earth for so many hours it makes the earth very warm.

There is another important reason why summer days are so warm. In summer the sun climbs high in the sky and shines almost straight down. This makes the earth very warm.

Summer days are long and hot.

Now, what about the next step in our cycle? Summer days follow one after the other and then it is autumn. As autumn comes the days get shorter and shorter and the sun does not climb so high in the sky. It is no longer nearly over your head at noon. In autumn the days begin to get cooler.

Autumn days are shorter and cooler than summer days.

Now comes winter, the next season in the cycle. The days keep getting shorter and shorter. In winter it is probably still dark when you wake up in the morning, and after supper it is already too dark for you to play outside. In winter at noon when you look for the sun it is much lower in the sky. Short days and the low sun make winter days cold.

Winter days are short and cold.

But in March the days begin to get warmer again. The sun begins to climb higher in the sky, and it stays in the sky for more hours. You can guess what happens to the days. You know that they begin to get warmer and warmer. Spring comes.

Spring days are longer and warmer than winter days.

After spring comes and goes, summer comes, and our cycle of the seasons begins again.

There is still another way to discover what happens during the cycle of the seasons. You can watch the living things around you. Plants and animals will show you. They change as the seasons change. As the cycle of the seasons pass, plants and animals go through cycles too.

EARLY IN THE SPRING, SEE WHAT PLANTS SHOW YOU. LOOK CAREFULLY AT BUDS AND BRANCHES.

WILLOW

FORSYTHIA

The buds are the beginning of a cycle of leaves and flowers. Bring some of the twigs and branches indoors and put the stems in water. Try to find willow and forsythia twigs, and apple and cherry tree twigs. Use a toothpick to open some of the buds and use your magnifying glass to look inside. Keep watching some of the buds as the days pass and you will discover how the cycle begins.

CHERRY

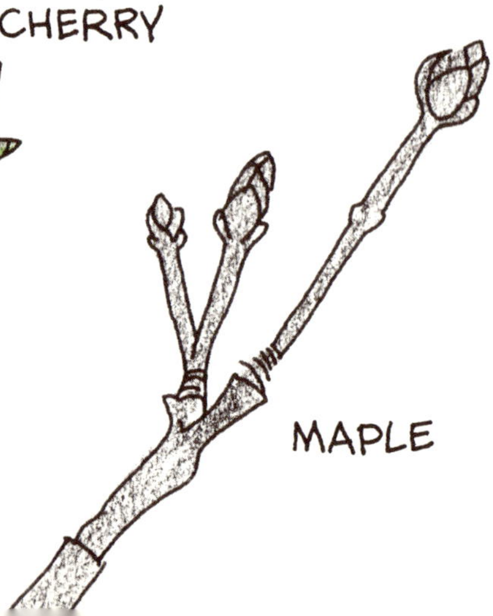

MAPLE

ARE PLANTS THAT ARE DIFFERENT ABOVE THE GROUND DIFFERENT UNDER THE GROUND TOO?

Dig down into the ground in fields and gardens and woods and search for seeds and bulbs. They begin to swell up and grow in spring. Soon tiny plants will poke up into the warm sunshine. All about you plants are beginning their cycles. Keep searching during spring and you will discover that grass, bushes, trees, flowers, garden vegetables, and farm crops are beginning to grow. All of this happens because the sun comes up earlier and climbs higher and stays up longer.

Keep up your search as summer comes. Some of the blossoms have come and gone. The leaves are now full-grown. Watch and you will see plants change in many other ways as summer goes by.

After a few months of summer, autumn comes. Take another trip to the fields and gardens and woods and see what you can discover. Search for dried seedpods on the plants you saw in spring.

Some plants die in autumn. Only their seeds live through the winter. They stay in the ground and are ready for another cycle when spring comes. The leaves change to yellow, orange, or red in autumn. After a while they fall to the ground and turn brown. Now if you look carefully at the twigs and branches, you will find the tiny buds ready for the spring season. When they open in spring a new cycle begins.

Animals have cycles too. In the spring you will probably see birds that you have not seen since last spring. Last autumn they gathered in flocks and flew south. Some flew only a short distance. Some flew hundreds of miles. We call this flight south in autumn and back north in spring *migrating*. Now they are back and you will see them build nests. After their nests are made they lay eggs and the cycle of the birds begins.

SOME SQUIRRELS' NESTS LOOK LIKE BIG, MESSY BIRDS' NESTS.

In spring other animals begin to raise their families too. Many insect eggs that were laid in the autumn begin to hatch, and these six-legged animals begin their cycles. Squirrels and other fur animals have babies. Spring is the season when many animals start their families.

Summer is a feeding and growing time for many animals. Baby birds and other animals grow larger and leave their homes.

Autumn is a busy time for many animals. Chipmunks and squirrels and some other kinds of animals store food for winter. Their fur gets heavier. Many birds migrate south. Some animals eat and get fat, then find a place underground or in hollow logs where they can pass the winter season. They *hibernate*. That means that they lie quietly through the winter months. Woodchucks, frogs and toads, and some other animals hibernate.

Winter is a quiet time for many animals. Some are hibernating. Some stay in sheltered places when the weather is very cold and only come out to find food on pleasant warmer days.

Then spring comes and the cycles begin again.

FROGS HIBERNATE IN THE MUD AT THE BOTTOMS OF PONDS.

# The Water Cycle

Lie on your back on a summer day and look up at the sky. See any clouds? Watch them. Sometimes they get smaller and lighter until they disappear. Sometimes they get larger and darker until they fill the sky. They hide the sun. Soon you may feel a sprinkle on your face. Then more drops fall. Better run for your house. It's raining!

Where does rain come from? What makes it rain? What are clouds made of? Is it true that water from a mud puddle may later rain from the sky? To answer these questions you must know about another cycle—the water cycle. Like the other cycles, it starts in one place and finally comes back to the same place again.

An example of the water cycle is going on in this picture. Can you tell what is happening?

There are two parts to the water cycle. First the water changes from a liquid to a gas. We say that it *evaporates.* Then it changes from a gas back into a liquid. We say that it *condenses.* Water evaporates and condenses and evaporates and condenses, over and over again.

Water evaporates from everything that is wet. From mud puddles in the street, from the wet ground, from dew, from seas and oceans and rivers and creeks, from wet clothes on the line, from watercolors and wet bathing suits. It evaporates from the leaves of grass and other plants too.

Water is always changing to water vapor. You can't see water vapor in the air. You are looking right through some of it as you read this sentence.

When the air is warm it can hold more water vapor than it can when it is cold. On a warm day the air may have a great deal of water vapor in it. But if the air gets cool enough, it can't hold the water vapor. Then you had better put on your raincoat or put up your umbrella because water may start to come out of the air. The water vapor condenses. It is raining. The water comes back to the ground and makes wet pavements and cars and fields and meadows and everything else that is outdoors. It makes puddles and falls into ponds and seas and oceans and rivers and creeks.

Now what happens? The next time it rains, see if you can discover what happens to the water that falls.

Some of it begins to flow down toward a lower place. If you are in the city, you can watch it run down the gutter into the sewer, where it goes under the city streets and runs into a creek or river. If you live in the country, you can watch water run downhill into creeks and rivers. Some of it keeps flowing down until it reaches the ocean. Oceans and seas get the water that runs from the lands around them.

Some of the water soaks into the ground. The roots of plants take it in. It goes up the stems to the leaves. Some of it stays in the plant, but some of it goes back into the air from the leaves of the plants.

Some of the water that falls on the pavements and meadows and fields and cars and other places soon begins to disappear. Some of it evaporates into the air. Water evaporates from everything that is wet.

Soon the air has lots of water vapor in it. If it gets cooler, you know what happens. The air can't hold all this water vapor, so some of it changes back into tiny drops of water. This makes a cloud like the cloud that comes out of a teakettle.

Not all of the water that comes out of the air and back to the earth makes rain. If you walk on the grass on an early cool morning, you may get your shoes wet from water that has come from the air. This is called *dew*. If the air is cold enough, the water may condense into ice crystals. This is called *frost*.

If you are watching clouds when the air is very cold, some of the water vapor condenses into ice crystals. You can see the clouds get larger and darker. If the water vapor keeps on condensing into ice crystals, these crystals start to get heavy. Some begin to fall to the earth. They are white and cold. At first there are only a few flakes. Then there are more and more. It is snowing!

If you like to slide downhill or make snowballs or a snowman, you are glad to see the snowflakes make a white cover on the ground.

You can draw a picture to show what happens in the water cycle. Start by drawing the things you see around you. The houses, streets, trees, grass, and sidewalks. Draw a creek and a river and a pond and a cloud or two. Use colored arrows to show what is happening to the water. Use one color to show what happens when the rain falls on the earth. Then put the sun in the picture and use another color to draw arrows showing what is happening to the water now. What name will you give your drawing? Does it look like the one on page 48?

The water cycle is very important to you. Suppose water just evaporated into the air and never condensed. Suppose water never evaporated? Suppose it never rained? Or snowed? Then what would happen?

# Some Plant Cycles

While the day and night cycle and the seasons cycle and the water cycle are going on, plants go through their cycle that brings flowers and fruits and vegetables.

A good way to see a plant cycle is to start with some flower seeds. Zinnia seeds are good. Spring is the best time to begin. Take a good look at the zinnia seeds with a magnifying glass. They look dry and dead. They only need some warmth from the sun and some rain from the sky, the soil and you. Just plant them, following the directions on the package.

If the ground is dry, sprinkle it. Soon things will begin to happen. The seeds will begin to sprout.

Watch for the first tiny green sprouts. After five or six days dig up one of the tiny plants and look at it with your magnifying glass. What can you see? Any roots? Any stem? Any leaves?

ZINNIAS

Watch the plants as spring changes to summer. Do they grow taller? Grow more leaves? Grow more branches? Grow buds? Grow flowers? What else happens? Perhaps you will make a record of what happens.

Don't pick all the flowers. Leave some of them on the plant and observe them. The colors fade. The flowers dry up. Leave a flower until summer changes to autumn, then pick it. Hold it over a piece of white paper and take it apart. What do you see? Use your magnifying glass. Have you seen something like this before? Can you count them? Save the seeds all winter, and when spring comes you can start the cycle over again. If you keep several flowers, you will have many more seeds than you had in the package when you started.

MARIGOLDS

NASTURTIUMS

Each year zinnias begin their cycles from seeds. Such plants are called *annuals*. Their cycle from seed to seed takes only one year. Many common plants are annuals. Zinnias, marigolds, and nasturtiums are annuals. So are the other flowers on this page. Look in a seed catalogue and you will find others.

PETUNIAS

CORNFLOWERS

COSMOS

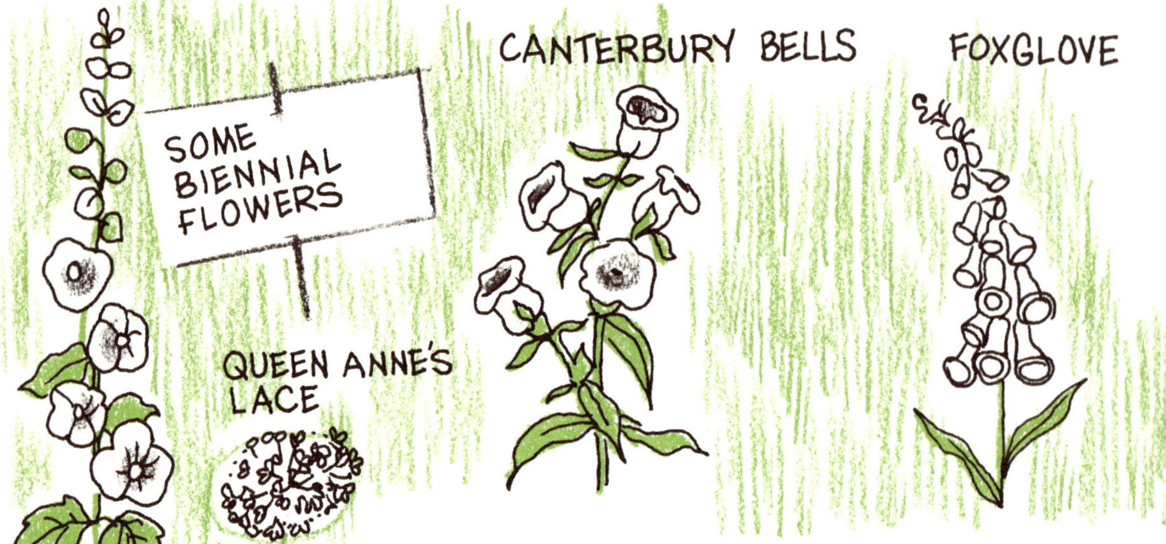

Not all flowering plants are annuals. Buy a package of foxglove or Canterbury bells or hollyhock seeds. Use your magnifying glass and make a drawing of the seeds so you will remember what they look like. Plant them in good soil. Observe them. They grow, they get more leaves and branches just as annuals do, but you won't see any flower buds or flowers. Fall comes, then winter. The leaves may turn brown and fall off, but the plant roots do not die. Wait until spring and see what happens.

In spring new leaves grow and the plant gets larger. Now watch for flower buds and then flowers. Let some of the flowers stay on the plant until they are brown and dry. Find the seeds. Do they look like the seeds you planted? How are these plants different from the annuals? These plants are called *biennials.* Their life cycle takes two years. The earth revolves around the sun twice before biennials complete their cycle. Look in the seed catalogue and you will find the names of other biennial plants.

Look in a garden to find roses or irises or peonies. They are not annuals or biennials. They are called *perennials*. They may lose their leaves in the wintertime, but their roots stay alive in the ground. When spring comes they grow new shoots and the buds open on branches and new leaves and flowers appear. They may live for many years and grow seeds year after year. A great number of garden flowers are perennials. So are trees and shrubs.

Some plants do not have flowers, and of course they have no seeds. Do you know any such plants? Ferns have no flowers nor seeds. Neither do mushrooms nor molds nor other plants that are not green, but they go through cycles just the same.

Have you ever seen a plant that has no stems or roots or leaves or flowers or seeds and is not even green? You can grow it on a piece of bread. Moisten the bread and lay it on a dish and shake a dusty cloth over it. Cover the bread with another dish that you can see through. Keep the bread moist and warm. Soon tiny plants will cover the bread.

This plant is mold. Look at it with your magnifying glass and you will see tiny black balls. There are many very small spores inside them. Mold plants do not have seeds. They have spores instead of seeds. When the balls break open the tiny spores come out. Spores are very light. A small breeze can carry them so there are mold spores almost everywhere. There were some in the dust you had on the cloth.

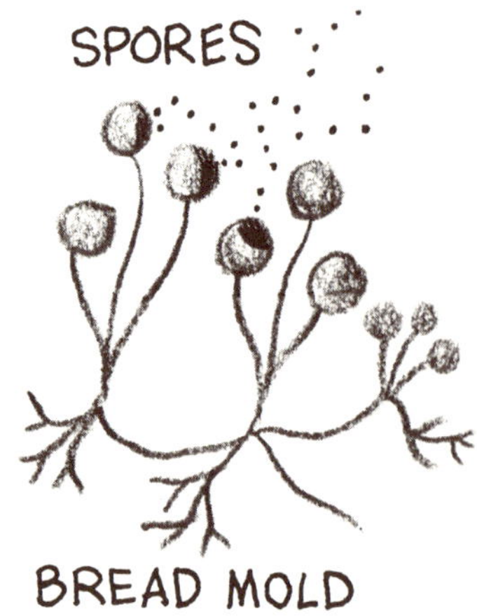


Spores begin the mold cycle just as seeds begin the cycle of other plants.

THERE ARE MANY KINDS OF MOLD. IF YOU LOOK AT THEM WITH YOUR MAGNIFYING GLASS, SOME OF THEM LOOK LIKE THIS

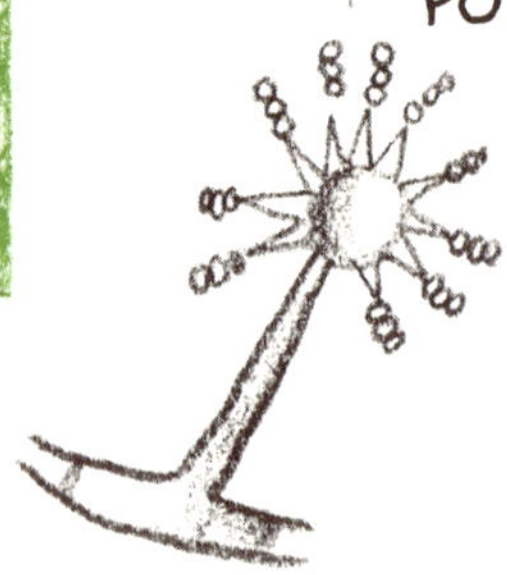


All plants have life cycles. While the earth is revolving around the sun and the seasons are following each other, you will see them if you look around. Some cycles take only a short time. Some take longer.

EGGS ARE TINY

# Some Animal Cycles

If you want to see an animal cycle, find some caterpillars. They are part of an insect cycle. You can find them on many kinds of plants. Look on carrot or parsley plants and you may find green-and-black-striped caterpillars. They are easy to feed and watch. Put them in a box and feed them the leaves. They will eat and grow and eat and grow. One day while you are watching you may see a strange thing happen. You may see one caterpillar climb out of its skin, and there underneath is a new skin, green with black stripes just like the old skin. This is called *molting*. The caterpillars will keep eating and growing and molting, and then one day some of them may be missing. Search and you may find some strange, brown, beautiful objects hanging in the box. Do you know what they are?

AT FIRST THE CATERPILLARS ARE SMALL AND SPINY WITH YELLOW POLKADOTS

THEN THEY LOOK LIKE THIS

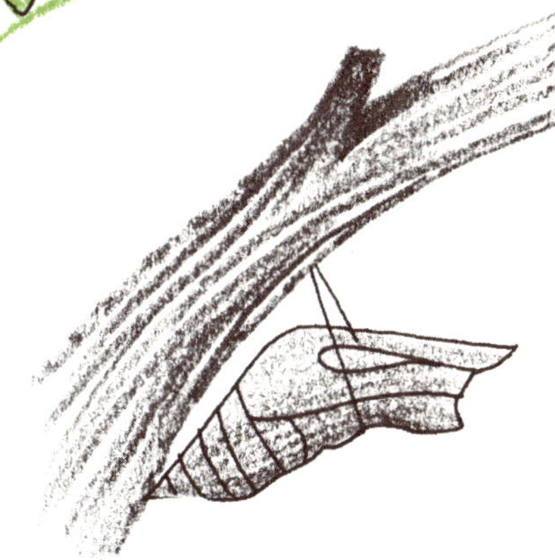

AND THEN (IF YOU PUT A DEAD BRANCH IN THE BOX) LIKE THIS

The caterpillar has changed into a *chrysalis.* Watch the chrysalis. Look at it with your magnifying glass.

Something is happening, but it is happening inside the chrysalis where you can't see it. The animal is changing. One day you may lift the cover and a beautiful butterfly will flutter and fly out. It is a black swallowtail butterfly.

Open the door and let it fly outside. It will probably fly back to the carrot or parsley plants in the garden. There it will find other black swallowtail butterflies. If the butterfly is a female, it will mate with a male butterfly. Before long it will lay tiny eggs on the leaves of the plants. Will the eggs hatch into butterflies? No, they hatch into tiny caterpillars. They are so small you will hardly notice them, but they begin to eat and grow and eat and grow. The cycle of the black swallowtail has started again.

You can describe the life cycle of butterflies in four words—egg, caterpillar (another name is *larva,* or larvae if there are more than one), chrysalis (another name is *pupa),* and butterfly (another name is *adult).* These are the four steps, or stages, in the life cycle.

There are many, many kinds of butterflies. Look for their larvae on milkweed, cabbage, clover, and many other plants. Now you know how to discover their cycles.

Moths have life cycles just as butterflies do. Instead of making a chrysalis, a moth spins a cocoon. The pupa is inside the cocoon.

There are many kinds of moths. Look for their larvae on tomato plants or on the leaves of trees and bushes like lilac, poplar, willow, apple, and cherry.

Moths and butterflies and many other insects have four stages in their life cycles, and many of their cycles do not take very long. Some kinds spend the winter in cocoons. Some pass the winter in the egg stage.

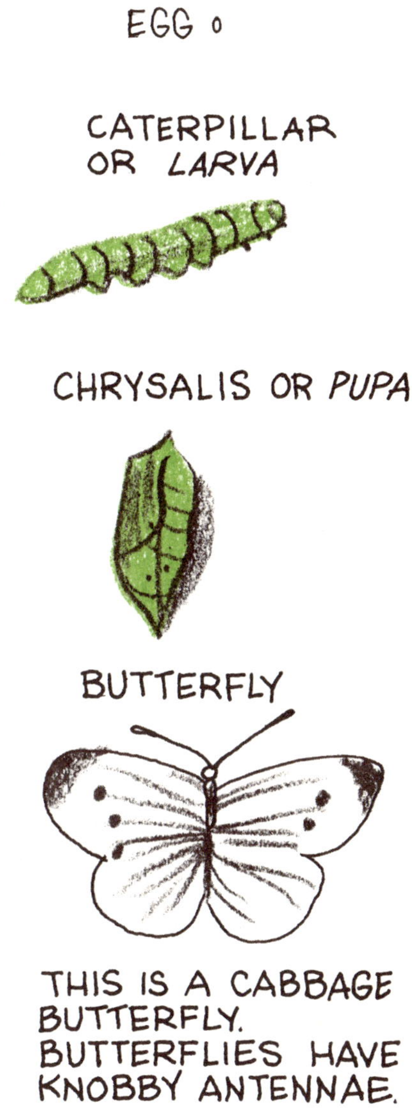


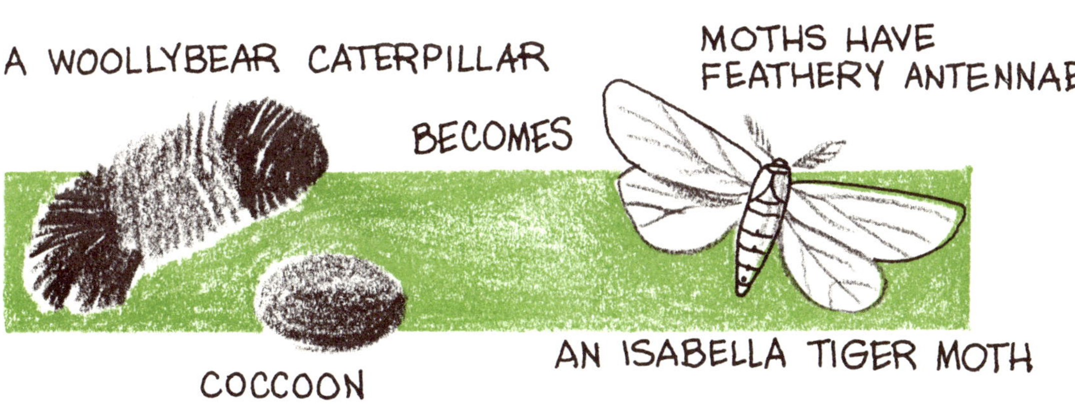

The life cycle of the next insect you will read about is full of surprises. It can only be seen once in a while. In the months of May and June of the year 1970 a strange thing happened in many parts of the eastern United States. Hundreds of tiny holes began to appear in the lawns and fields, and strange-looking small creatures climbed out. A few days later there was such a loud buzzing and screeching noise coming from somewhere on the branches of the trees that you could hardly hear yourself think. After a few weeks the noise stopped. A week or so later the tips of the tree branches began to turn brown. On some trees the leaves at the tip of nearly every branch were dead. A few weeks later these tips began to fall to the ground.

These creatures had not been seen for 17 years! That was the last time anyone had heard such a buzzing and screeching noise or seen the brown tips on tree branches. That was the last time that the insect called the 17-year locust (whose real name is *cicada)* had been heard from.

Much of the life cycle of the 17-year cicada is not easy to see because it happens in the ground. It takes 17 years!

The holes in the ground are about the size of a pencil or sometimes a little larger. They are made by the strange-looking creatures as they dig themselves out of the ground. If you look at the right spot at the right time, you can see them poking and wiggling up and out of the holes. This is the stage of the cicada called a *nymph*.

The nymphs crawl to a nearby tree trunk or plant stem and climb up. Soon they stop and then a remarkable thing happens. The hard outer shell of the nymph splits down the back and out crawls the adult cicada.

Some of the adults are males. Some are females. The male cicadas are the ones that make the noise.

The males and females mate and soon the male cicadas die. Then the loud buzzing and screeching noise stops. The female insects crawl way out to the ends of the branches. There they make slits around the twigs and lay their eggs in them. Then the female cicada dies too.

The cuts in the ends of the twigs make the twigs weak, the twigs bend, and the leaves wilt and turn brown. Wind and rain break the twigs off and they fall to the ground.

The eggs that the female laid hatch into tiny nymphs. Those that hatch from the eggs on the twigs that fell to the ground don't have far to go to begin to dig down into the ground. Those that are left on the tree fall to the ground and begin their digging. Down, down they go—sometimes as far as several feet.

The nymphs live a very quiet life for years and years. They suck sap from roots in the ground. They grow slowly. They molt. Seventeen years go by.

In the year 1987 people will begin to notice holes in the ground and hear the noises and read in the newspaper that this is the year for the 17-year locust. You are probably thinking, "That's a long life cycle." And it is. It is the longest life cycle of any insect. How many cycles does the earth make around the sun while the cicada is making its cycle?

Not all kinds of cicadas take 17 years to complete their life cycles. Some take 13 years and some take less. Other animals go through cycles too. When you hear frogs and toads croaking in creeks and ponds you know that it is spring. When you hear the croaking you can be almost sure that you will find eggs.

They are laid in the water where the frogs and toads live. The eggs are the beginning of a very interesting life cycle.

It is easy to tell toads' eggs from frogs' eggs. Frogs' eggs are laid in a lump like a ball, and toads' eggs are laid in long strings. Look for a place where the water is quiet. Look carefully, for the eggs appear almost like the place where they are laid.

Scoop up a bunch of eggs with a pail. Take plenty of pond water and some green plants and pond scum with them. You need only a few eggs. Leave the others to hatch, because frogs and toads are very helpful animals. They eat many harmful insects.

Put the eggs, plants, and water into a gallon jar or an aquarium. Don't set them in the hot sunlight. You may need to add more pond water now and then as you watch the cycle. Do you know what will happen to the eggs?

Start your discovering by using your magnifying glass to look at the eggs. How would you describe them? You may want to make a picture record of your discoveries. Draw the egg. Draw the shape of the tiny tadpole when you first see it. Draw pictures that show changes. Label the pictures and put down the dates. How long does it take for the eggs to hatch? What do

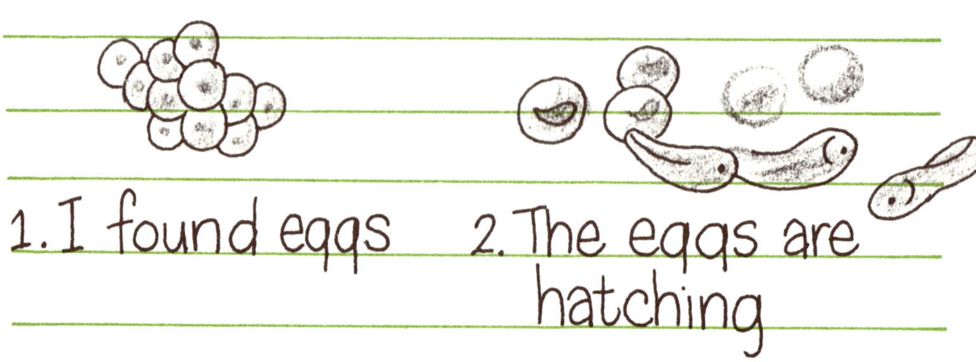

the tiny tadpoles look like? How do they swim? Do they have legs? Do the front legs or the hind legs come first? Do both of the front legs grow at the same time? What happens to the tail?

At first the eggs are black and round. Each black egg has a jelly-like material around it. Soon the eggs change shape. They begin to look like tadpoles. Then you may see the tiny animals wiggle and jerk a little. Keep watching and you will see them wiggle out of the jelly and begin to swim. The frog cycle has started.

Keep watching the tadpoles. Can you see a mouth? How do they swim? Some of them will hold on to the plants that you brought from the pond. Use your magnifying glass and you may see the gills on either side of the head. The tiny tadpole gets air from the water by using these gills. Tadpoles feed on the pond scum and other plants you brought with the water. Try to feed them tiny pieces of lettuce or bits of the yolk of hard-boiled eggs. Try other things. Take out food that the tadpoles do not eat.

SPRING PEEPERS ARE SMALL.
THEIR LIFE CYCLE TAKES ONLY
A FEW WEEKS.

BULLFROGS ARE BIG.
THEIR LIFE CYCLE TAKES
MORE THAN A YEAR.

Some tadpoles go through their life cycles much faster than others. One kind called the spring peeper takes only a few weeks or so. Some kinds take a whole year and some take even longer. That is why you may find some good-sized tadpoles when you are looking for eggs. They began their life cycle last year or perhaps the year before, when they hatched from eggs. They probably stayed in the mud in the bottom of the pond when winter came. Large tadpoles like these do very well in an aquarium. You can watch them change to adult frogs.

All animals have cycles. Keep watching animals that you see around you. What can you discover about the life cycles of squirrels? Of cats? Of dogs?

# Cycles in an Aquarium

A good aquarium has clean sand in the bottom with plants growing in it. It has fish, some snails, and maybe a tadpole or two. A cycle goes on in an aquarium that you can't see. It's a cycle of two gases. One is oxygen. The other is carbon dioxide. Both of these gases are in the air around you, and they are in the water too, but you can't see them in either place. Here are a few things you need to know before you can understand the cycle of these gases.

Fish need oxygen to live just as you do. You get the oxygen you need from the air you breathe. Take a deep breath. Air goes into your nose and down into your lungs. Then your body gets the oxygen it needs.

But fish cannot breathe air as you do. They do not have lungs. Instead they have gills. These gills can take the oxygen out of the air in the water. You cannot see the air in water. But fill a glass with water and let it sit for a few hours. You will see bubbles on the inside of the glass. These bubbles are air. They came out of the water.

The gills of the fish are inside their bodies where you can't see them. Watch the fish and you will see that they seem always to be drinking. They are not drinking. They are opening and closing their mouths. That water goes in through the fish's mouth and comes out through the gills. The gills take the oxygen out of the water.

Now you know where fish get the oxygen they need. But you are probably thinking that fish will soon use up all the oxygen that is in the air in the water. They would, except that something keeps putting more oxygen into the water. That's where the cycle comes in. Can you imagine what living things in an aquarium can keep putting oxygen back into the water? The plants do. That's why you need plants in an aquarium.

Plants need oxygen just as animals do. But green plants are different from animals in one important way. They can make food. They need two things in their food-making and both of them are in the aquarium. One is water, and there is plenty of that. The other is carbon dioxide. Now where do you suppose that comes from? It comes from the fish. They give it off into the water through their gills when they breathe.

FISH TAKE IN OXYGEN AND GIVE OFF CARBON DIOXIDE THROUGH THEIR GILLS.

PLANTS TAKE IN CARBON DIOXIDE AND GIVE OFF OXYGEN.

$O^2$ IS THE CHEMICAL SYMBOL FOR OXYGEN

$CO^2$ IS THE CHEMICAL SYMBOL FOR CARBON DIOXIDE

Try making a picture of what happens in the oxygen and carbon dioxide cycle. Draw a fish. Draw some plants.

What label will you put with the arrow pointing into the leaves? What label for the one coming out? What label will you put with the arrow pointing into the gills of the fish? What label pointing out from the gills? Now draw lines that connect the two $O_2$ labels and the two $CO_2$ labels. Look on page 48 and you will find a drawing of the $O_2$ and $CO_2$ cycle. Is it like yours?

What would happen in an aquarium if there were no green plants? No animals?

The $O_2$ and $CO_2$ cycle happens all around you. It goes on wherever plants and animals grow under water and wherever they grow on land too. The leaves of trees, grass, bushes, vegetables, flowers, and all green plants keep putting $O_2$ into the air. Birds, frogs, squirrels, people, and other living animals keep using the $O_2$ and putting some $CO_2$ back into the air.

Don't get the idea that plants breathe $CO_2$. They don't. They need $O_2$ just as animals do. But you remember that plants need $CO_2$ in their leaves to use in manufacturing food. It is from this food-making that the $O_2$ comes that they give off into the air.

Did you ever see a water cycle in an aquarium? What have you seen under the glass cover when it got cool? Where did these drops come from? What will happen to them when it gets warm again?

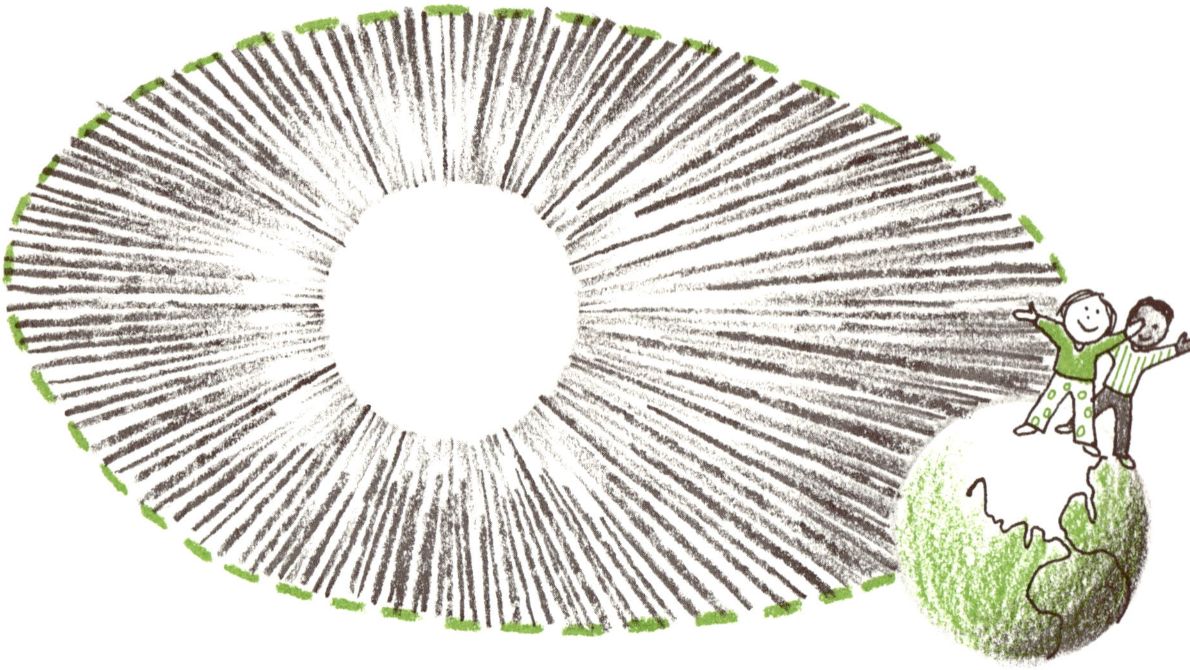

# More Cycles

While you have been reading this book the second and minute hands of your watch have been making cycles. You have been traveling more than 66,000 miles an hour on the earth as it revolves around the sun. The day and night cycle has been going on. Water has been evaporating and condensing. It has fallen on plants and animals that are making cycle changes as the seasons pass.

There are many other cycles. The moon changes from new moon to first quarter, to full moon, to third quarter, and back to new moon again every month. Watch it and you will see.

What happens to the leaves of trees when they fall to the ground in autumn? Walk in the woods and dig down under the leaves and see if you can discover what is happening to them. The cycle of leaves is a very important one to you.

A gas called nitrogen makes a cycle that is so important to you that you couldn't live without it. Can you find out about it?

There are many other cycles that you will learn about as the earth revolves and the seasons pass and you grow older. Look for them around you. Do you make a cycle yourself?

DOES YOUR WATER CYCLE PICTURE LOOK SOMETHING LIKE THIS?

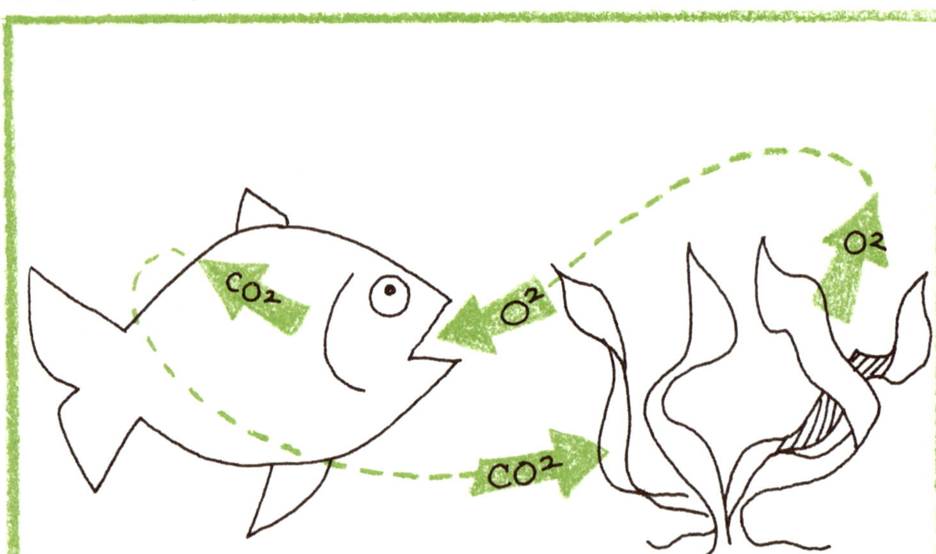


DOES YOUR OXYGEN AND CARBON DIOXIDE CYCLE PICTURE LOOK SOMETHING LIKE THIS?